Tucholsky Wagner Zola Scott Sydow Schlegel
Turgenev Wallace Fonatne Freud
Twain Walther von der Vogelweide Fouqué Friedrich II. von Preußen
Weber Freiligrath Frey
Fechner Fichte Weiße Rose von Fallersleben Kant Ernst Richthofen Frommel
Hölderlin
Engels Fielding Eichendorff Tacitus Dumas
Fehrs Faber Flaubert Eliasberg Ebner Eschenbach
Feuerbach Maximilian I. von Habsburg Fock Eliot Zweig
Ewald Vergil
Goethe London
Elisabeth von Österreich
Mendelssohn Balzac Shakespeare Dostojewski Ganghofer
Lichtenberg Rathenau Doyle Gjellerup
Trackl Stevenson Tolstoi Hambruch
Mommsen Thoma Lenz Hanrieder Droste-Hülshoff
Dach Verne von Arnim Hägele Hauff Humboldt
Reuter Rousseau Hagen Hauptmann Gautier
Karrillon Garschin Defoe Hebbel Baudelaire
Damaschke Descartes
Hegel Kussmaul Herder
Wolfram von Eschenbach Darwin Dickens Schopenhauer Rilke George
Bronner Melville Grimm Jerome Bebel
Campe Horváth Aristoteles Proust
Bismarck Vigny Barlach Voltaire Federer Herodot
Gengenbach Heine
Storm Casanova Tersteegen Gilm Grillparzer Georgy
Chamberlain Lessing Langbein Gryphius
Brentano Claudius Schiller Lafontaine
Strachwitz Bellamy Schilling Kralik Iffland Sokrates
Katharina II. von Rußland Gerstäcker Raabe Gibbon Tschechow
Löns Hesse Hoffmann Gogol Wilde Gleim Vulpius
Luther Heym Hofmannsthal Klee Hölty Morgenstern Goedicke
Roth Heyse Klopstock Kleist
Luxemburg Puschkin Homer Mörike Musil
Machiavelli La Roche Horaz
Navarra Aurel Musset Kierkegaard Kraft Kraus Moltke
Nestroy Marie de France Lamprecht Kind Kirchhoff Hugo
Nietzsche Nansen Laotse Ipsen Liebknecht
Marx Lassalle Gorki Klett Leibniz Ringelnatz
von Ossietzky May vom Stein Lawrence Irving
Petalozzi Platon Pückler Michelangelo Knigge Kafka
Sachs Poe Liebermann Kock
de Sade Praetorius Mistral Zetkin Korolenko

The publishing house tradition has created the series **TREDITION CLASSICS**. It contains classical literature works from over two thousand years. Most of these titles have been out of print and off the bookstore shelves for decades.

The book series is intended to preserve the cultural legacy and to promote the timeless works of classical literature. As a reader of a **TREDITION CLASSICS** book, the reader supports the mission to save many of the amazing works of world literature from oblivion.

The symbol of **TREDITION CLASSICS** is Johannes Gutenberg (1400 – 1468), the inventor of movable type printing.

With the series, tradition intends to make thousands of international literature classics available in printed format again – worldwide.

All books are available at book retailers worldwide in paperback and in hardcover. For more information please visit: www.tredition.com

tradition was established in 2006 by Sandra Latusseck and Soenke Schulz. Based in Hamburg, Germany, tradition offers publishing solutions to authors and publishing houses, combined with worldwide distribution of printed and digital book content. tradition is uniquely positioned to enable authors and publishing houses to create books on their own terms and without conventional manufacturing risks.

For more information please visit: www.tredition.com

Handbook on Japanning: 2nd Edition For Ironware, Tinware, Wood, Etc. With Sections on Tinplating and Galvanizing

William N. Brown

Imprint

This book is part of the TREDITION CLASSICS series.

Author: William N. Brown
Cover design: toepferschumann, Berlin (Germany)

Publisher: tredition GmbH, Hamburg (Germany)
ISBN: 978-3-8491-6591-8

www.tredition.com
www.tredition.de

Copyright:
The content of this book is sourced from the public domain.

The intention of the TREDITION CLASSICS series is to make world literature in the public domain available in printed format. Literary enthusiasts and organizations worldwide have scanned and digitally edited the original texts. tredition has subsequently formatted and redesigned the content into a modern reading layout. Therefore, we cannot guarantee the exact reproduction of the original format of a particular historic edition. Please also note that no modifications have been made to the spelling, therefore it may differ from the orthography used today.

CONTENTS.

SECTION I.

INTRODUCTION.

   Priming or Preparing the Surface to be Japanned

   The First Stage in the Japanning of Wood or of Leather without a Priming

SECTION II.

JAPAN GROUNDS.

   White Japan Grounds

   Blue Japan Grounds

   Scarlet Japan Ground

   Red Japan Ground

   Bright Pale Yellow Grounds

   Green Japan Grounds

   Orange-Coloured Grounds

   Purple Grounds

   Black Grounds

   Common Black Japan Grounds on Metal

   Tortoise-shell Ground

   Painting Japan Work

   Varnishing Japan Work

SECTION III.

JAPANNING OR ENAMELLING METALS.

　Enamelling Bedstead Frames and similar large pieces

　Japanning Tin, such as Tea-trays and similar goods

　Enamelling Old Work

SECTION IV.

THE ENAMELLING AND JAPANNING STOVE — PIGMENTS SUITABLE FOR JAPANNING WITH NATURAL LACQUER — MODERN METHODS OF JAPANNING WITH NATURAL JAPANESE LACQUER.

　Appliances and Apparatus used in Japanning and Enamelling

　Modern Japanning and Enamelling Stoves

　Stoves heated by direct fire

　Stoves heated by hot-water pipes

　Pigments suitable for Japanning with Natural Lacquer

　White Pigments

　Red Pigments

　Blue Pigment

　Yellow Pigments

　Green Pigment

　Black Pigment

　Methods of Application

　Modern Methods of Japanning and Enamelling with Natural Japanese Lacquer

SECTION V.

COLOURS FOR POLISHED BRASS. — MISCELLANEOUS.

- Painting on Zinc or on Galvanized Iron
- Bronzing Compositions
- Golden Varnish for Metal
- Carriage Varnish
- Metal Polishes
- Black Paints
- Black Stain for Iron
- Varnishes for Ironwork

SECTION VI.

PROCESSES FOR TIN-PLATING.

- Amalgam Process
- Immersion Process
- Battery Process
- Weigler's Process
- Hern's Process

SECTION VII.

GALVANIZING.

HANDBOOK ON JAPANNING.

SECTION I.

INTRODUCTION.

Japanning, as it is generally understood in Great Britain, is the art of covering paper, wood, or metal with a more or less thick coating of brilliant varnish, and hardening the same by baking it in an oven at a suitable heat. It originated in Japan—hence its name—where the natives use a natural varnish or lacquer which flows from a certain kind of tree, and which on its issuing from the plant is of a creamy tint, but becomes black on exposure to the air. It is mainly with the application of "japan" to metallic surfaces that we are concerned in these pages. Japanning may be said to occupy a position midway between painting and porcelain enamelling, and a japanned surface differs from an ordinary painted surface in being far more brilliant, smoother, harder, and more durable, and also in retaining its gloss permanently, in not being easily injured by hot water or by being placed near a fire; while real good japanning is characterised by great lustre and adhesiveness to the metal to which it has been applied, and its non-liability to chipping—a fault which, as a rule, stamps the common article.

If the English process of japanning be more simple and produces a less durable, a less costly coating than the Japanese method, yet its practice is not so injurious to the health. Indeed, it is a moot point in how far the Japanese themselves now utilize their classical process, as the coat of natural japan on all the articles exhibited at the recent Vienna exhibition as being coated with the natural lacquer, when recovered after six months' immersion in sea water through the sinking of the ship, was destroyed, although it stood perfectly well on the articles of some age. In the English method, where necessary, a priming or undercoat is employed. It is customary to fill up any uneven surface, any minute holes or pores, and to render the surface to be japanned uniformly smooth. But such an undercoat or priming is not always applied, the coloured varnish or a proper japan ground being applied directly on the surface to be japanned. Formerly this surface usually, if not always, received a priming coat, and it does so still where the surface is coarse, uneven, rough, and porous. But where the surface is impervious and smooth, as in the case of metallic surfaces, a priming coat is not applied. It is also unnecessary to apply such a coat in the case of smooth, compact, grained wood. The reason for using this coating is that it effects a considerable saving in the quantity of varnish used, and because the matter of which the priming is composed renders the surface of the body to be varnished uniform, and fills up all pores, cracks, and other inequalities, and by its use it is easy after rubbing and water polishing to produce an even surface on which to apply the varnish. The previous application of this undercoat was thus an advantage in the case of coarse, uneven surfaces that it formed a first and sort of obligatory initial stage in the process of japanning. This initial coating is still applied in many instances. But it has its drawbacks, and these drawbacks are incidental to the nature of the priming coat which consists of size and whiting. The coats or layers of japan proper, that is of varnish and pigment applied over such a priming coat, will be continually liable to crack or peel off with any violent shock, and will not last nearly so long as articles japanned with the same materials and altogether in the same way but without the undercoat. This defect may be readily perceived by comparing goods that have been in use for some time in the japanning of which an undercoat has been applied with similar goods in which no such previous coat has been given. Provided a good japan varnish and appropriate pigments have been used and the japanning well executed, the coats of japan applied without a priming never peel or crack or are in any way damaged except by violence or shock, or that caused by continual ordinary wear and tear caused by such constant rubbing as will wear away the surface of the japan. But japan coats applied with a priming coat crack and fly off in flakes at the slightest concussion, at any knock or fall, more especially at the edges. Those Birmingham manufacturers who were the first to practise japanning only on metals on which there was no need for a priming coat did not of course adopt such a practice. Moreover, they found it equally unnecessary in the case of papier-mâché and some other goods. Hence Birmingham japanned goods wear better than those goods which receive a priming previous to japanning.

Priming or Preparing the Surface to be Japanned.

The usual priming, where one is applied, consists of Paris white (levigated whiting) made into a thin paste with size. The size should be of a consistency between the common double size and glue, and mixed with as much Paris white as will give it a good body so that it will hide the surface on which it is applied. But in particular work glovers' or parchment size instead of common size is used, and this is still further improved by the addition of one-third of isinglass, and if the coat be not applied too thickly it will be much less liable to peel or crack. The surface should be previously prepared for this priming by being well cleaned and by being brushed over with hot size diluted with two-thirds of water, that is provided the size be of the usual strength. The priming is then evenly and uniformly applied with a brush and left to dry. On a fairly even surface two coats of priming properly applied should suffice. But if it will not take a proper water polish, owing to the uneven surface not being effectually filled up, one or more additional coats must be applied. Previous to the last coat being applied, the surface should be smoothed by fine glass paper. When the last coat of priming is dry the water polish is applied. This is done by passing a fine wet rag or moistened sponge over the surface until the whole appears uniformly smooth and even. The priming is now complete and the surface ready to take the japan ground or the coloured varnish.

The First Stage in the Japanning of Wood or of Leather Without a Priming.

[The leather is first securely stretched on a frame or board.] In this case, that is when no priming coat is previously applied, the best way to prepare the surface is to apply three coats of coarse varnish (1 lb. seed-lac, 1 lb rosin to 1 gallon methylated spirit, dissolve and filter). This varnish, like all others formed from methylated spirits, must be applied in a warm place and all dampness should be avoided, for either cold or moisture chills it and thus prevents it taking proper hold of the surface on which it is applied. When the work is prepared thus, or by the priming made of size and whiting already described, the japan proper is itself applied.

SECTION II.

JAPAN GROUNDS.

The japan ground properly so called consists of the varnish and pigment where the whole surface is to be of one simple colour, or of the varnish, with or without pigment, on which some painting or other form of decoration is afterwards to be applied. It is best to form this ground with the desired pigment incorporated with shellac varnish, except in the case of a white japan ground which requires special treatment, or when great brilliancy is a desideratum and other methods must be adopted. The shellac varnish for the japan ground is best prepared as follows: shellac $1^1/_4$ lb., methylated spirits 1 gallon. Dissolve in a well-corked vessel in a warm place and with frequent shaking. After two or three days the shellac will be dissolved. It is then recommended to filter the solution through a flannel bag, and when all that will come through freely has done so the varnish should be run into a proper sized vessel and kept carefully corked for use. The bag may then be squeezed with the hand till the remainder of the fluid varnish is forced through it, and this if fairly clear may be used for rough purposes or added to the next batch. Pigments of any nature whatever may be used with the shellac varnish to give the desired tint to the ground, and where necessary they may be mixed together to form any compound colour, such as blue and yellow to form green. The pigments used for japan grounds should all be previously ground very smooth in spirits of turpentine, so smooth that the paste does not grate between the two thumb nails, and then only are they mixed with the varnish. This mixture of pigment and varnish vehicle should then be spread over the surface to be japanned very carefully and very evenly with a camel-hair brush. As metals do not require a priming coat of size and whiting, the japan ground may be applied to metallic surfaces forthwith without any preliminary treatment except thorough cleansing, except in the cases specially referred to further on. On metallic surfaces three to four coats are applied, and in the interval between each coat the articles must be stoved in an oven heated to from 250° to 300° F.

White Japan Grounds.

The formation of a perfectly white japan ground and of the first degree of hardness has always been difficult to attain in the art of japanning, as there are few or no substances that can be so dissolved as to form a very hard varnish coat without being so darkened in the process as to quite degrade or spoil the whiteness of the colour. The following process, however, is said to give a composition which yields a very near approach to a perfect white ground: Take flake white or white lead washed and ground up with the sixth of its weight of starch and then dried, temper it properly for spreading with mastic varnish made thus: Take 5 oz. of mastic in powder and put it into a proper vessel with 1 lb. of spirits of turpentine; let them boil at a gentle heat till the mastic be dissolved, and, if there appear to be any turbidity, strain off the solution through flannel. Apply this intimate and homogeneous mixture on the body to be japanned, the surface of which has been suitably prepared either with or without the priming, then varnish it over with five or six coats of the following varnish: Provide any quantity of the best seed-lac and pick out of it all the clearest and whitest grains, take of this seed-lac $1/2$ lb. and of gum anime $3/4$ lb., pulverize the mixture to a coarse powder and dissolve in a gallon of methylated spirits and strain off the clear varnish. The seed-lac will give a slight tint to this varnish, but it cannot be omitted where the japanned surface must be hard, though where a softer surface will serve the purpose the proportion of seed-lac may be diminished and a little turpentine oleo-resin added to the gum anime to take off the brittleness. A very good varnish entirely free from brittleness may, it is said, be formed by dissolving gum anime in old nut or poppy oil, which must be made to boil gently when the gum is put into it. After being diluted with turps the white ground may be applied in this varnish, and then a coat or two of the varnish itself may be applied over it. These coats, however, take a long time to dry, and, owing to its softer nature, this japanned surface is more readily injured than that yielded by the shellac varnish.

According to Mr. Dickson, "the old way of making a cream enamel for stoving (a white was supposed to be impossible) was to mix ordinary tub white lead with the polishing copal varnish and to add a modicum of blue to neutralize the yellow tinge, stove same in about 170°F. and then polish as before described". "This," continues Mr. Dickson, "would at the best produce but a very pale blue enamel or a cream. It was afterwards made with flake white or dry white lead ground in turps only and mixed with the polishing copal varnish with the addition of tints as required, by which means a white of any required character could be produced."

Blue Japan Grounds.

Authorities state that these may be formed from bright Prussian blue or verditer glazed over with Prussian blue or of smalt. By bright Prussian blue possibly a genuine Prussian blue toned down to a sky blue with white lead is meant, and by verditer the variety known as refiners' blue verditer, and as to smalt it must not be forgotten that it changes its colour in artificial light. Be that as it may, the pigment may be mixed with the shellac varnish according to the instructions already given, but as the shellac will somewhat injure the tone of the pigment by imparting a yellow tinge to it where a bright true blue is required, the directions already given as regards white grounds must be carried out.

Scarlet Japan Ground.

Vermilion is the best pigment to use for a scarlet japan ground, and its effect will be greatly enhanced by glazing it over with carmine or fine lake. If, however, the highest degree of brightness be required the white varnish must be used. Vermilion must be stoved at a very gentle heat.

Red Japan Ground.

The basis of this japan ground is made up with madder lake ground in oil of turpentine, this constitutes the first ground; when this is perfectly dry a second coat of lake and white in copal varnish is applied, and the last coat is made up of lake in a mixture of copal varnish and turpentine varnish.

Bright Pale Yellow Grounds.

Orpiment or King's yellow may be used, and the effect is enhanced by dissolving powdered turmeric root in the methylated spirits from which the upper or polishing coat is made, which methylated spirits must be strained from off the dregs before the seed-lac is added to it to form the varnish. The seed-lac varnish is not so injurious to yellow pigments as it is to the tone of some other pigments, because, being tinged a reddish yellow, it does little more than intensify or deepen the tone of the pigment.

Green Japan Grounds.

Green japan grounds are produced by mixing Prussian blue or distilled verdigris with orpiment, and the effect is said to be extremely brilliant by applying them on a ground of leaf gold. Any of them may be used with good seed-lac varnish, for reasons already given. Equal parts by weight of rosin, precipitated rosinate of copper, and coal-tar solvent naphtha will give a varnish which, when suitably thinned and the coats stoved at a heat below 212° F., will give a green japan second to none as a finishing coat as regards purity of tone at least. To harden it and render it more elastic half of the rosin might be replaced by equal weights of a copal soluble in solvent naphtha and boiled linseed oil, so that the mixture would stand thus: rosinate of copper 1 lb., rosin $1/2$ lb., boiled oil $1/4$ lb., hard resin (copal) $1/4$ lb., solvent naphtha 1 lb. When heated to a high temperature this rosinate of copper varnish yields a magnificent ruby bronze coloration, especially on glass. Verdigris dissolves in turpentine, and successful attempts might be made to make a green japan varnish from it on the lines indicated for rosinate of copper.

Orange-coloured Grounds.

Orange-coloured grounds may be formed by mixing vermilion or red lead with King's yellow, or orange lake or red orpiment (? realgar) will make a brighter orange ground than can be produced by any mixture.

Purple Grounds.

Purple grounds may be produced by the admixture of lake or vermilion with Prussian blue. They may be treated as the other coloured grounds as regards the varnish vehicle.

Black Grounds.

Black grounds may be formed either from lamp black or ivory black, but ivory black is preferable to lamp black, and possibly carbon black or gas black to either. These may be always applied with the shellac varnish as a vehicle, and their upper or polishing coats may consist of common seed-lac varnish. But the best quality of ivory black ground in the best super black japan yields, after suitable stoving, a very excellent black indeed, the purity of tone of which may be improved by adding a little blue in the grinding.

Common Black Japan Grounds On Metal.

Common black japan grounds on metal by means of heat are procured in the following manner: The surface to be japanned must be coated over with drying oil, and when it is moderately dry must be put into a stove of such heat as will change the oil black without burning it. The stove should not be too hot when the oil is put into it nor the heat increased too fast, either which error would make it blister, but the slower the heat is increased and the longer it is continued, provided it be restrained within a due degree, the harder will be the coat of japan. This kind of japan requires no polish, having received from the heat, when properly regulated, a sufficiently bright surface.

Tortoise-Shell Ground.

This beautiful ground, produced by heat, is valued not only for its hardness and its capacity to stand a heat greater than that of boiling water, but also for its fine appearance. It is made by means of a varnish prepared thus: Take one gallon of good linseed oil and half a pound of umber, boil them together until the oil becomes very brown and thick, strain it then through a coarse cloth and set it again to boil, in which state it must be continued until it acquires a consistency resembling that of pitch; it will then be fit for use. Having thus prepared the varnish, clean well the surface which is to be japanned; then apply vermilion ground in shellac varnish or with drying oil, very thinly diluted with oil of turpentine, on the places intended to imitate the more transparent parts of the tortoise-shell. When the vermilion is dry, brush the whole over with the black varnish thinned to the right consistency with oil of turpentine. When set and firm put the work into a stove where it may undergo a very strong heat, which must be continued a considerable time, for three weeks or even a month so much the better. This ground may be decorated with painting and gilding in the same way as any other varnished surface, which had best be done after the ground has been hardened, but it is well to give a second annealing at a very gentle heat after it has been finished. A very good black japan may be made by mixing a little japan gold size with ivory or lamp-black, this will develop a good gloss without requiring to be varnished afterwards.

Painting Japan Work.

Japan work should be painted with real "enamel paints," that is with paints actually ground in varnish, and in that case all pigments may be used and the peculiar disadvantages, which attend several pigments with respect to oil or water, cease with this class of vehicle, for they are secured by it when properly handled from the least danger of changing or fading. The preparation of pigments for this purpose consists in bringing them to a due state of fineness by grinding them on a stone with turpentine. The best varnish for binding and preserving the pigments is shellac. This, when judiciously handled, gives such a firmness and hardness to the work that, if it be afterwards further secured with a moderately thick coat of seed-lac varnish, it will be almost as hard and durable as glass. The method of painting in varnish is, however, far more tedious than with an oil or water vehicle. It is, therefore, now very usual in japan work for the sake of dispatch, and in some cases in order to be able to use the pencil (brush) more freely, to apply the colours in an oil vehicle well diluted with turps. This oil (or japanners' gold size) may be made thus: Take 1 lb. of linseed oil and 4 oz. of gum anime, set the oil in a proper vessel and then add the gum anime powder, stirring it well until the whole is mixed with the oil. Let the mixture continue to boil until it appears of a thick consistence, then strain the whole through a coarse cloth and keep it for use. The pigments are also sometimes applied in a gum-water vehicle, but work so done, it has been urged, is not nearly so durable as that done in varnish or oil. However, those who formerly condemned the practice of japanning water-coloured decorations allowed that amateurs, who practised japanning for their amusement only and thus might not find it convenient to stock the necessary preparations for the other methods, might paint with water-colours. If the pigments are ground in an aqueous vehicle of strong isinglass size and honey instead of gum water the work would not be much inferior to that executed with other vehicles. Water-colours are sometimes applied on a ground of gold after the style of other paintings, and sometimes so as to produce an embossed effect. The pigments in this style of painting are ground in a vehicle of isinglass size corrected with honey or sugar-candy. The body with which the embossed work is raised is best formed of strong gum water thickened to a proper consistency with armenian bole and whiting in equal parts, which, being laid on in the proper figures and repaired when dry, may be then painted with the intended pigments in the vehicle of isinglass size or in the general manner with shellac varnish. As to the comparative value of pigments ground in water and ground in oil, that is between oil-colours and water-colours in enamelling and japanning, there seems to have been a change of opinion for some time back, especially as regards the enamelling of slate. The marbling of slate (to be enamelled) in water-colours is a process which Mr. Dickson says well repays study. It is greatly developed in France and Germany. The process is a quick one and the pigments are said to stand well and to maintain their pristine hue, yet if many strikingly natural effects result from the use of this process, its use has not spread in Great Britain, being confined wholly and solely to the marbling of slate (except in the case of wall-paper which is water-marbled in a somewhat similar way).

"In painting in oil-colour," says Mr. Dickson, "the craftsman trusts largely to his badger-hair brush to produce his effects of softness and marbly appearance; but in painting in water-colours, this softness, depth, and marbly appearance are produced mostly by the colour placed upon the surface, and left entirely untouched by badger or any other brush. The colour drying quickly, does not allow much time for working, and when dry it cannot be touched without spoiling the whole of the work. The difference first of all between painting in water and in oil colour, is that a peculiar grain exists with painting in water that it is absolutely impossible to get in oil. The charm of a marble is, I think, its translucency as much as its beautiful colour; it is to that translucency (for in marble fixed we have no transparency) that it owes its softness of effect, which makes marble of such decorative value. This translucency can only be obtained by thin glazes of colour, by which means each succeeding glaze only partly covers the previous one, the character of the marble being thus produced. This is done sometimes in oil-colour in a marvellous manner, but even the best of oil-painting in marble cannot stand the comparison of water-colour, and it is only by comparison that any accurate judgment can be formed of any work. The production of marbles in water-colour has a depth, softness, and stoniness that defies oil-painting, and in some cases will defy detection unless by an expert of marbles. It may be that first of all the materials employed are more in keeping with the real material, as no oil enters into the composition of real marble, and by using the medium of water we thus start better, but the real secret is that by using water as a medium the colours take an entirely different effect. In painting in water-colour greys of any tint or strength can be obtained suitable for the production of a marble of greyish ground, by pure white, tinted as required, being applied of different thicknesses of colour, all the modulations of tone being obtained by the difference in the thickness of the colour applied."

Varnishing Japan Work.

Varnishing is the last and the finishing process in japanning. It consists in (1) applying, and (2) polishing the outer coats of varnish, which are equally necessary whether the plain japan ground be painted on or not. This is best done in a general way with common seed-lac varnish, except on those occasions where other methods have been shown to be more expedient, and the same reasons, which decide as to the propriety of using the different varnishes as regards the colours of the ground, hold equally with those of the painting, for where brightness is a material point and a tinge of yellow would injure it, seed-lac must give way to the whiter resins; but where hardness and tenacity are essential it must be adhered to, and where both are necessary a mixed varnish must be used. This mixed varnish should be made from the picked seed-lac as directed in the case of the white japan grounds. The common seed-lac varnish may be made thus: Take $1^1/_2$ lb. of seed-lac and wash it well in several waters, then dry it and powder it coarsely and put it with a gallon of methylated spirits into a Bohemian glass flask so that it be not more than two-thirds full. Shake the mixture well together and place the flask in a gentle heat till the seed-lac appears to be dissolved, the shaking being in the meantime repeated as often as may be convenient; then pour off all the clear and strain the remainder through a coarse cloth. The varnish so prepared must be kept for use in a well-corked glass vessel. The whiter seed-lac varnishes are used in the same manner as the common, except as regards the substances used in polishing, which, where a pure white or the greater clearness or purity of other pigments is in question, should be itself white, while the browner sorts of polishing dust, as being cheaper and doing their business with greater dispatch, may be used in other cases. The pieces of work to be varnished should be placed near the fire or in a warm room and made perfectly dry, and then the varnish may be applied with a flat camel-hair brush made for the purpose. This must be done very rapidly, but with great care; the same place should not be passed twice over in laying on one coat if it can possibly be avoided. The best way of proceeding is to begin in the middle and pass the brush to one end, then with another stroke from the middle pass it to the other end, taking care that before each stroke the brush be well supplied with varnish; when one coat is dry another must be laid over it in like manner, and this must be continued five or six times. If on trial there be not a sufficient thickness of varnish to bear the polish without laying bare the painting or ground colour underneath more varnish must be applied. When a sufficient number of coats of varnish is so applied the work is fit to be polished, which must be done in common work by rubbing it with a piece of cloth or felt dipped in tripoli or finely ground pumice-stone. But towards the end of the rubbing a little oil of any kind must be used with the powder, and when the work appears sufficiently bright and glossy it should be well rubbed with the oil alone to clean it from the powder and to give it a still greater lustre. In the case of white grounds, instead of the tripoli, fine putty or whiting should be used, but they should be washed over to prevent the danger of damaging the work from any sand or any other gritty matter that may happen to be mixed with them. It greatly improves all kinds of japan work to harden the varnish by means of heat, which, in every degree that can be applied short of what would burn or calcine the matter, tends to give it a firm and strong texture where metals form the body; therefore a very hot stove may be used, and the stoving may be continued for a considerable time, especially if the heat be gradually increased. But where wood or papier-mâché is in question, heat must be applied with great caution.

SECTION III.

JAPANNING OR ENAMELLING METALS.

In japanning metals, all good work of which should be stoved, they have to be first thoroughly cleaned, and then the japan ground applied with a badger or camel-hair brush or other means, very carefully and evenly. Metals usually require from three to five coats, and between each application must be dried in an oven heated from 250° to 300° F.—about 270° being the average. It has already been seen that the best grounds for japanning are formed of shellac varnish, the necessary pigments for colouring being added thereto, being mixed with the shellac varnish after they have been ground into a high degree of smoothness and fineness in spirits of turpentine. In japanning it is best to have the oven at rather a lower temperature, increasing the heat after the work has been placed in the oven. When a sufficient number of coats have been laid on—which will usually be two only—the work must be polished by means of a piece of cloth or felt dipped in tripoli or finely powdered pumice-stone. For white grounds fine putty powder or whiting must be employed, a final coat being afterwards given, and the work stoved again. The last coat of all is one of varnish. And here, as a preliminary remark, it is advisable that all enamels and japans should be purchased ready-made, as any attempt to make such is almost sure to end in disaster, while, owing to the fact that such are only required for small jobs; it would involve too much trouble and would not pay. It is for this reason that few japan recipes are given, as, although many are available, they do not always turn out as suitable for the purpose as could be desired, in addition to which the ready-made articles can be purchased at a very reasonable price and are much better prepared. The operator should procure his enamels a shade or two lighter than he desires to see in the finished article, allowing the chemical action due to the stoving to tone the colours down. Another necessity is to keep the enamel thoroughly well mixed by well stirring it every time it is used, as if this is not done the actual colouring matter is apt to sink to the bottom, the ultimate result being that streaky work is produced in consequence of this indifferent mixing of the enamelling materials.

It is hardly necessary to state that all japanning or enamelling work must be done in a room or shop absolutely free from dust or dirt, and as far away as possible from any window or other opening leading to the open air, for two reasons—one being that the draught therefrom may cool the oven or stove, and the other that the air may convey particles of dust into the enamelling shop. In fact, it cannot be too much impressed upon the workmen that one of the primary secrets of successful enamelling is absolute cleanliness; consequently all precautions must be taken to ensure that the enamel is perfectly free from grit and dust, and it must be so kept by frequent straining through fine muslin, flannel, or similar material. The work having been thoroughly cleaned and freed from all grease and other foreign matter, it must be suspended or held immediately over the pan elsewhere referred to, and the enamel poured on with an ordinary iron ladle, or covered by means of the brush. When it has been permitted to drain thoroughly, the work should be hung on the hooks on the rods in the oven as seen in the explanatory sketch, care being observed that no portion of the work is in such a position that any superfluous enamel cannot easily drain off—in other words, the work must lie or hang that it is always, as it were, on the slant. Always bear in mind when shutting the oven door to do so gently, as if a slam is indulged in all the gas jets will be blown out, and an explosion would probably result.

Should the job in hand be a large one, it will be found as well to get a cheaper enamel for the first coat, but if the work is only a small job, it will not be necessary to have more than one enamel, of which a couple of coats at least will be required. When the first coat has thoroughly dried and hardened, the surface will have to be thoroughly rubbed till it is perfectly smooth with tripoli powder and fine pumice-stone, and afterwards hand-polished with rotten-stone and putty powder. And here it may be remarked that the finer the surface is got up with emery powder and other polishing agents the better will be the enamelling and ultimate finish. The rubbing down being finished, another coat of enamel must be applied and the work baked as before, care being always taken to keep the enamel in a sufficiently fluid condition as to enable it to flow and run off the work freely. It can easily be thinned with a little paraffin. A third coat will frequently be advisable, as it improves the finish.

In enamelling cycles, it is well to hang the front forks crown uppermost when they are undergoing the final baking, and it is advisable to bear in mind that wheels require an enamel that will stove at a lower temperature than is called for for other parts of the machine. Some japanners advocate the fluid being put on with camel-or badger-hair brushes, and for the best descriptions of work, final coats, and such like, I agree with them; but this is a detail which can be left to the operator's own fancy, the class of work, etc.; but I would remind him that applying enamel with a brush requires much care and a certain amount of "knack". It is something like successful lacquering in brasswork — it looks very simple, but is not. Each succeeding coat of japan gives a more uniform and glossy surface, and for this reason it may, in some cases, be necessary to repeat the operation no fewer than half a dozen times, the final coat being generally a layer of clear varnish only, to add to the lustre.

Care must be taken for light-coloured japans or enamels not to have the temperature sufficiently high to scorch, or the surface will be discoloured, as they require a lower temperature for fixing than the dark japans, which, provided the article is not likely to be injured by the heat, are usually dried at a somewhat high temperature. The preceding instructions apply only to the best descriptions of work.

When pouring enamel by means of the ladle over pieces of work, do not agitate the liquid too much — at the same time taking care to keep it well mixed — so as to form air bubbles, as this will cause trouble, and in pouring over the work do it with an easy and gentle and not too hurried a motion. In japanning curved pieces, such as mud-guards, etc., in hanging up the work in the oven see that the liquid does not run to extremities and there form ugly blots or blotches of enamel.

When white or other light tones are used for japanning they are mixed with japanners' varnish, and these require more careful heating in the oven or stove than darker tints or brown or black.

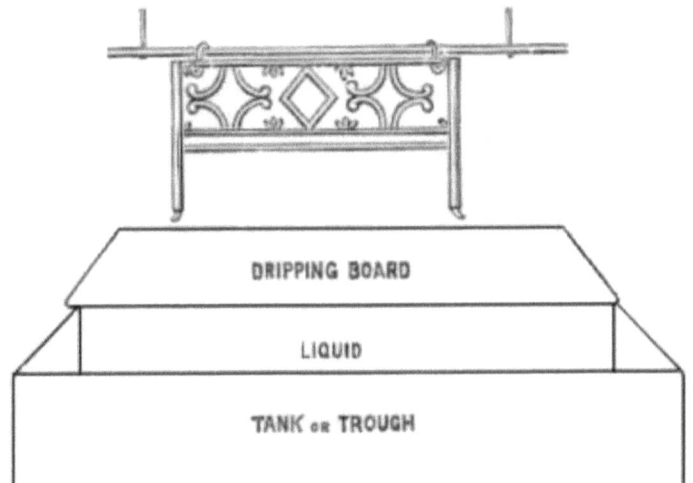

Fig. 1.—Trough for Dipping Bedstead Frames and other Large Work.

Enamelling Bedstead Frames and Similar Large Pieces.

At Fig. 1 is shown a trough in which large pieces, such as bedsteads, bicycle frames, etc., are dipped or immersed. For the first-mentioned class of work such high finish is not required as for bicycles, and consequently the enamel need not be applied with a brush, nor will it be necessary to rub down the work between each coat, but instead the pieces can be literally dipped in the tank of liquid, then allowed to drain on to the dripping-board—the superfluous enamel thus finding its way back into the trough or tank, the dripped articles being afterwards placed in the oven to harden. The trough must be of sufficient dimensions to allow the pieces of work to be completely immersed, and the dripping-board should be set at an angle of about 45°.

Bedstead frames will never require more than two coats and the commoner class of goods only one. I would not advise the tradesman in a small way of business to go to the expense of a trough, etc., as it calls for much more room than is ordinarily available, but if he has the necessary plant for bicycle work he can, of course, do an occasional job of the other kind.

Japanning Tin, such as Tea-trays and Similar Goods.

For japanning sheet-iron articles, which are really tin goods, such as tea-trays and similar things, first scour them well with a piece of sandstone, which will effectually remove all the scales and make the surface quite smooth. Then give the metal a coating of vegetable black, which must be mixed with super black japan varnish, thinned with turps, and well strained. Only a small quantity of this varnish is necessary, as it will dry dead. The article must then be placed in the stove to harden at a temperature of 212° F., there to remain for from ten to twelve hours. When taken out of the stove, the articles must be allowed to get cold, after which they must be given a coat of super black japan, which, if necessary, must be thinned with turps, a stiff, short bristle brush being employed, and the varnish put on sparingly, so that it will not "run" when it gets warm. Two coats of this varnish on top of the vegetable black coating are usually sufficient, when done properly, but a third coating much improves the work, and from ten to twelve hours' hardening will be necessary between each coating. The small lumps which will be more or less certain to arise will require to be rubbed down between each application by a small and smooth piece of pumice-stone.

If it is desired to add gold or bronze bands or any kind of floral or other kind of fancy decorations, these are painted on, after the ground japanning has been done, in japanners' gold size, and then the gold leaf is applied, or the bronze or other metal powder is dusted on, after which the objects so treated are again placed in the stove, where they will not require to be kept near so long as for ordinary japanning. After they have been removed, the gilt or bronzed portions must be treated with a protecting coat of white spirit varnish. Transfers can be applied in the same way.

Tinned iron goods are the most largely japanned, and for these brown and black colours are principally employed. Both are obtained by the use of brown japan, the metal having a preliminary coating of black paint when black is required. Only one coating of brown japan is given to cheap goods, but for better articles two or more are applied. For these it is possible that a final dressing with pumice-stone, then with rotten-stone, and rubbed with a piece of felt or cloth, or even the palm of the hand, may be necessary, but as a rule not.

Large numbers of articles of the above description, such as tea-trays, tea-canisters, cash-boxes, coal-boxes, and similar goods, are japanned at Birmingham, and it is to such that the preceding instructions apply.

Enamelling Old Work.

In all cases of re-enamelling old work, it is absolutely necessary to remove all traces of the first enamelling, and if this has been well done in the first instance, it will prove no mean job. The best way to clean the work is to soak it in a strong "lye" of hot potash, when the softened enamel can be wiped or brushed off—this latter method being pursued in the more intricate and ungetatable portions of the work. New work, which has not been enamelled, can be treated in the same way for the removal of all grease, stains, finger-marks, etc., and too much attention cannot be paid to the initial preparation of the surface of the metal, to have it thoroughly even and smooth, as it adds so much to the ultimate finish and appearance of the work. Plenty of labour must be bestowed before the final coat, as any blemish will show through this finishing, and so mar what would otherwise be a highly satisfactory bit of work. In all kinds of bicycle work, whether new or old, the most satisfactory results are obtained by the application of at least two, and sometimes four or five, successive coats of good but thin enamel, as this will impart the necessary perfect coat, combined with durability, a high finish, and a good colour. A good enamel should be sufficiently hard, so as not to be scratched on the merest touch or rubbing. It will, of course, be understood that no solder-work must be put into the stove, or the pieces will separate. Should any of this work be discovered, the pieces must be taken apart, and then brazed together before being enamelled, and put in the stove.

SECTION IV.

THE ENAMELLING AND JAPANNING STOVE — PIGMENTS SUITABLE FOR JAPANNING WITH NATURAL LACQUER — MODERN METHODS OF JAPANNING WITH NATURAL JAPANESE LACQUER.

Appliances and Apparatus used in Japanning and Enamelling.

Besides the various enamels or japans and varnishes of various colourings and the stove, which will be found described and illustrated, together with the trough, in other pages, the worker will need some iron pots or cauldrons in which to boil the potash "lye" for the cleansing, more particularly, of old work, some iron ladles both for this work and for pouring the japan on the articles to be covered therewith, a few badger tools and brushes for small fine work, some hooks for the stove, a pair of pliers, a few bits of broom handle cut into short lengths and made taper, so as to fit into the tubes, etc., of bicycles and other work, so as to keep the hands as free from the japan as possible, some emery powder, pumice-stone powder, tripoli, putty powder, whiting, and a piece of felt or cloth. If he is also doing any common work, a stumpy brush of bristles and a soft leather will also be requisite, together with a file or two. These will about comprise the whole of the articles required, not very expensive, all of which will really not be required by a beginner.

Owing largely to the strides made in the cycle trade enamelling is stoved by means of gas, and of this a plentiful supply is necessary. Enamelling stoves may really be described as hot-air cupboards or ovens, and for a stove which will answer most requirements — say one of 6 feet by 6 feet by $3^1/_2$ feet — six rows of atmospheric burners will be necessary to heat it, while it will be also advisable to fix pipes of $1^1/_4$ inch internal diameter from the gas meter to the stove. The atmospheric burners can be made from the requisite number of pieces of $1^1/_4$-inch gas tube $3^1/_2$ feet in length, one end of each being stopped, and having $1/_3$-inch holes drilled therein at intervals of about 1 inch, the other end being left open for the insertion of ordinary $3/_8$-inch brass gas taps. Another plan preferred by some japanners is to have three rows of burners the full length of the stove, which, under some circumstances, due to structural conditions, will be found more suitable. Anyway, whatever the position of the stove, allowance must be made for a temperature up to 400° F. to be raised. In old-fashioned ovens the heat is applied by means of external flues, in which hot air or steam is circulated, but this system is generally unsatisfactory, the supply of heat having to be controlled by dampers or stop-cocks, and this has given place to the gas apparatus. Another simple form of oven, though not one which I shall recommend, is a species of sheet-iron box, which is encased by another and larger box of the same shape, so placed that from 2 to 3 inches of interspace exists between the two boxes. To this interspace heat is applied, and a flue will have to be affixed to this apparatus to carry off the vapours which arise from the enamel or japan. For amateur or intermittent jobbing work the oven illustrated in Figs. 2 and 3 is about as good as any, though to guard against fire it would be as well to have a course of brickwork beneath the oven, while if this is not possible on account of want of height, a sheet or so of zinc or iron will help to mitigate the danger. It is also advisable, if the apartment is a low-pitched one, to have a sheet of iron or zinc suspended by four corner chains from the ceiling in order to protect this from firing through the heat from the enamelling oven. Of course, it will be understood that every portion of the stove must be put together with rivets, no soldered work being permissible.

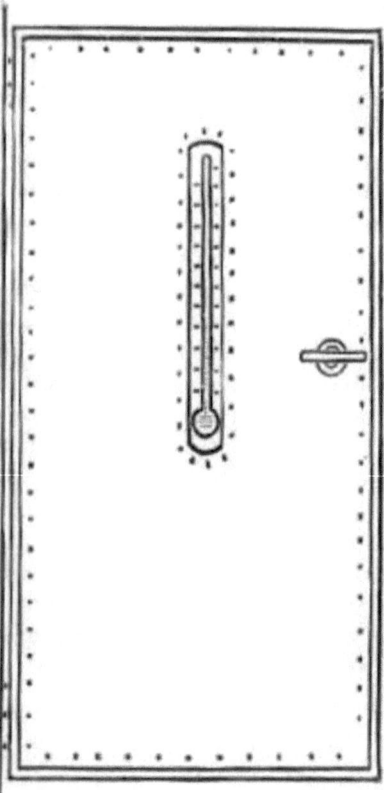

Fig. 2.—Door of Oven when Shut.

To those who wish to construct their own stove, it will be found that the framework can be shaped out of 1-inch angle iron, the panels or walls being constructed of sheet-iron of about 18 gauge, the whole being riveted together. The front will be occupied in its entire space by a door, which will require to be hung on strong iron hinges, and the framework of this door should be constructed of 1 inch by $1/4$ inch iron—a rather stouter material will really be no disadvantage—to which the sheet-iron plates must be riveted. In the centre of the door must be cut a slit, say $1^1/_2$ inches by 9 inches, which will require to be covered with mica or talc behind which must be placed the thermometer, so as it can be seen during the process of stoving, without the necessity of opening the door, which, of course, more or less cools the oven. And, by the way, this thermometer must register higher than the highest temperature the oven is capable of reaching. Above is shown a sketch of the stove, interior and exterior, which will give an idea of what a japanner's stove is like.

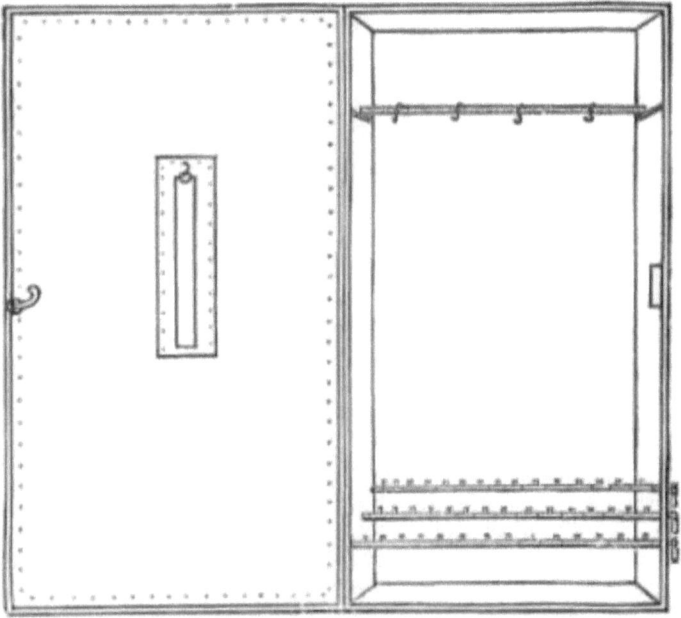

Fig. 3 — Showing Stove when Open, and Back of Door.

Inside the stove it will be necessary to fix rows of iron rods, some four inches from the top, from which to suspend the work, or angle-iron ledges can be used on which the rods or bars can be fixed, these arrangements being varied according to the particular description of work, individual fancy, or other circumstances. Large S hooks are about the handiest to use. A necessary adjunct of the stove is a pan, which can be made by any handy man or tinworker, which should be made to fit the bottom of the stove above the gas jets, it being arranged that it rests on two side ledges, or along some rods. One a couple of inches in depth will be found sufficient, and it will repay its cost in the saving of enamel, it being possible with its use to enamel a bicycle with as little as a gallon of enamel. Some workmen have the tray made with a couple of hinged side flaps, to turn over and cover up the pan when not in use, but this is a matter of fancy. Of course, they must always be covered up when not in use. For those who would prefer to use Bunsen burners, I show at Fig. 4 a sketch of the best to employ, these having three rows of holes in each.

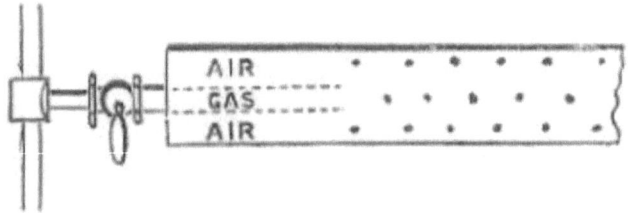

Fig. 4. – Bunsen Burner.

When brick ovens are employed they must be lined with sheet-iron, and in these very rare circumstances where gas is not available, the stove can be heated with coal or wood, which will, of course, involve a total alteration in the structural arrangements. I have not given the details here, as I do not think the necessity will ever arise for their use, and for the same reason I have refrained from giving the particulars for heating by steam and electricity, or the other methods which have been adopted by various workers, as there is no question but that a gas stove or oven, as described, is about the best and handiest for jobbers or amateurs.

Modern Japanning and Enamelling Stoves.

The modern japanning and enamelling stove consists of a compartment capable of being heated to any desired temperature, say 100° to 400° F., and at the same time, except as regards ventilation, capable of being hermetically sealed so as to prevent access of dust, soot, and dirt of all kinds to mar the beauty and lustre of the object being enamelled or japanned. Such a stove may be heated —

1. By a direct coal, coke, wood, peat, or gas fire (which surrounds the inner isolated chamber) (Fig. 5).

2. By heated air.

3. By steam or hot-water pipes, coils of which circulate round the interior of the stove or under the floor.

Such ovens may be either permanent, that is, built into masonry, or portable.

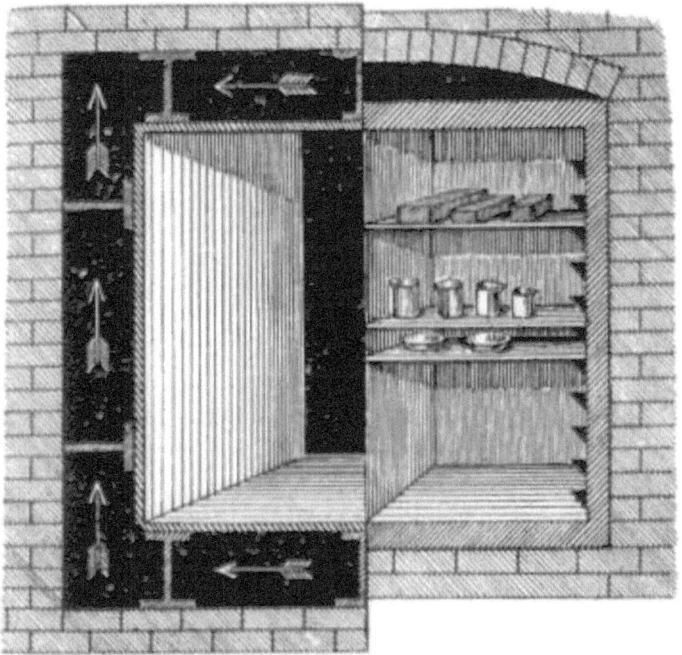

Fig. 5. — Greuzburg's Japanning Oven.

1. *Stoves heated by direct fire.* — These were, of course, the form in which japanning ovens were constructed somewhat after the style of a drying kiln. Fig. 5, Greuzburg's japanning oven heated on the

outside by hot gases from furnace. The oven is built into brickwork, and the hot gases circulate in the flues between the brickwork and the oven, and its erection and the arrangement of the heating flues are a bricklayer's job. Coke containing much sulphur is objectionable as a fuel for enamel stoves Mr. Dickson emphasizes this very forcibly. He says: "In the days when stoves were heated by coke furnaces, and the heat distributed by the flues, the principal trouble was the escape of fumes of sulphur which caused dire disaster to all the enamels by entering into their composition and preventing their ever drying, not to speak of hardening. I have known enamels to be in the stoves with heat to 270° for two and three days, and then be soft. The sulphur also caused the enamels to crack in a peculiar manner, much like a crocodile skin, and work so affected could never be made satisfactory, for here again we come back to the first principle, that if the foundation be not good, the superstructure can never be permanent. The enamels, being permeated with sulphur and other products from the coke, could never be made satisfactory, and the only way was to clean it all off. The other principal troubles are the blowing of the work in air bubbles, which is caused mainly by the heat being too suddenly applied to the articles, but these are very small matters to the experienced craftsman."

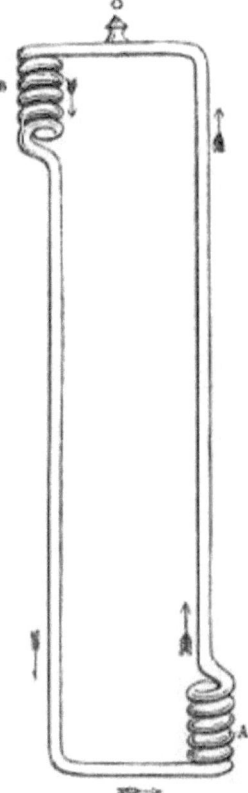

Fig. 6.

2. *Stoves heated by hot-water pipes.* — Let us first of all consider the principle on which these are constructed. In Perkins' apparatus for conveying heat through buildings by the circulation of water in small-bore hot-water pipes an endless tube or pipe is employed, the surface of which is occasionally increased by spiral or other turnings where the heat is to be given off or acquired: the annexed figure may serve to illustrate this principle; it represents a strong wrought-iron tube of about one inch diameter completely filled with water; the spiral A passes through a furnace where it is highly

heated, and the water is consequently put into motion in the direction of the arrows; the boiling of the water or formation of steam is prevented by the pressure, whence the necessity of the extreme perfection and strength of the tube. B represents a second coil which is supposed to be in an apartment where the heat is to be given out. C is a screw stopper by which the water may be occasionally replenished. By this form of apparatus the water may be heated to 300° or 400°, or even higher, so as occasionally to singe paper. A larger tube and lower temperature are, however, generally preferable. [1]

Fig. 7. — Enamelling Stove — in a Tin-plate Printing Factory — heated by Perkins' Hot-water Pipes.

The principle of Perkins' invention has, during the last eighty years, i.e. since the date of the invention in 1831, been very extensively applied not only for the heating of buildings of every description, but it has also been utilized for numerous industrial purposes which require an atmosphere heated up to 600° F. The principle lends itself specially to the design of apparatus for raising and maintaining heat evenly and uniformly, and also very economically for such purposes as enamelling, japanning, and lacquering.

The distinctive feature of this apparatus when applied to moderate temperatures lies in the adoption of a closed system of piping of small bore, a certain portion of which is wound into a coil and placed in a furnace situated in any convenient position outside the drying chamber or hot closet. The circulation is thus hermetically sealed and so proportioned that while a much higher temperature can be attained than is possible with a system of pipes open to the atmosphere, yet a certain and perfectly safe maximum cannot by any possibility be exceeded.

The efficiency of the apparatus increases within certain limits in proportion to the pressure employed, which fact explains the exceedingly economical results obtained, while the fact that, owing to the high temperature used, a small-bore pipe can be made more effective than the larger pipes used in any open system, accounts for the lower first cost of the Perkins' apparatus.

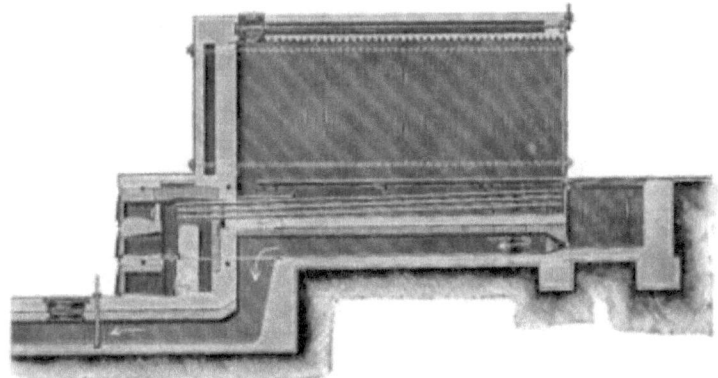

Fig. 8.—Japanning and Enamelling Oven Heated by Single Hot-water Pipes sealed at both ends with Furnace in Rear.

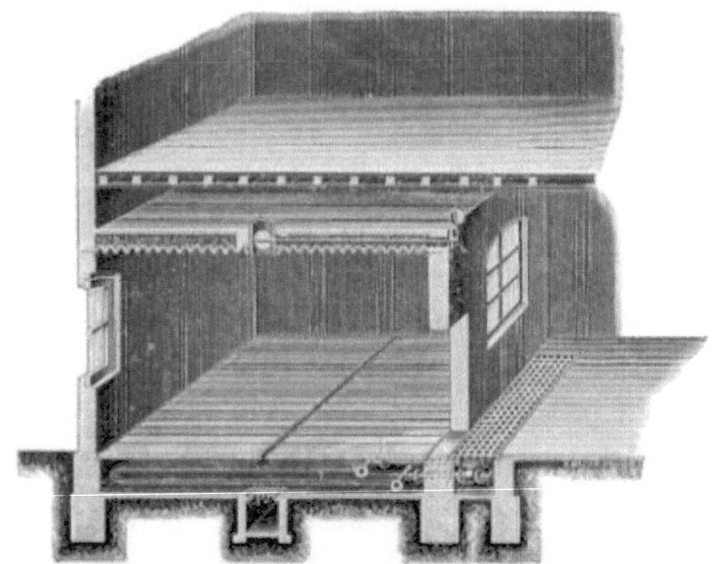

Fig. 9 – Japanning and Enamelling Oven For Bedstead, Ironmongery, Cash-box, and Lamp Factories.

Fig. 10.—Japanning and Enamelling Stove for parts of Sewing Machines.

It will be seen from the various illustrations that the articles to be treated are absolutely isolated from actual contact with the fire or the fire gases and other impurities which must be an objection to all methods of heating by means which are not of a purely mechanical nature. This principle not only recommends itself as scientifically correct and suited to the purpose in view, but is also a very simple and practical one. It affords the means of applying the heat at the point where it is required to do the work without unduly heating parts where heat is unnecessary; it secures absolute uniformity, perfect continuity, and the highest possible fuel economy.

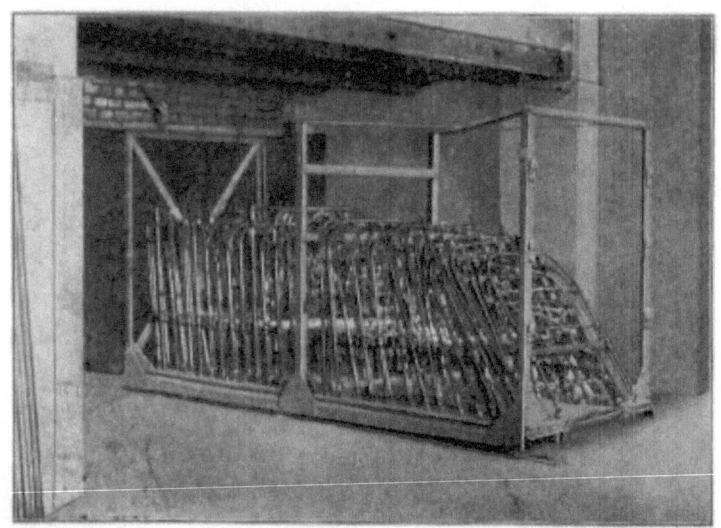

Fig. 11.—Japanning and Enamelling Stove for Iron-Bedsteads and Household Ironmongery with Truck on Rails.

Fig. 12 — Permanent Japanning and Enamelling Stove for Kitchen Utensils built in Masonry.

The nature of the work to be executed in the different classes and various sizes of stoves vary so greatly and indefinitely that only by careful attention to the special requirements of each case, on the part of the designers and constructors, is it possible to obtain the most satisfactory results.

The arrangement of fixing the pipes round the lower walls of the room in this form of stove is somewhat cumbersome, but in a roomy stove this slight drawback is not felt quite so much. However, it seems a good principle to leave every inch of internal space available for the goods to be enamelled or japanned, This principle is carried out to the letter in the other form of stoves described and illustrated in the sequel.

The figure shows a section through single chamber japanning and enamelling oven heated by hot-water pipes (steel) closed at both ends and partially filled with water which always remains sealed up therein, and never evaporates until the pipes require to be refilled.

This stove may be heated (1) by hot-water pipes (iron), (2) by super-heated water, (3) by steam, but only to 80° C. The different compartments may be heated to uniform or to different temperatures with hot water; the stoke-hole is at the side and thus quite separated from the stove proper.

The ovens must be on the ground floor, so that the super-heated steam from the basement may be available.

The great drawback to the use of gas for heating japanning and enamelling stoves is the great cost of coal gas.

Fig. 13.—Portable Gas Heated Japanning and Enamelling Stove fitted with Shelves, Thermometer, etc.

Pigments Suitable for Japanning with Natural Lacquer.

White Pigments.—Barium sulphate and bismuth oxychloride. These two are used for the white lacquer or as a body for coloured lacquers. When the lacquer is to be dried at a high temperature barium sulphate is preferable, but when it is dried at an ordinary temperature bismuth oxychloride is better. Since the lacquer is originally of a brown colour the white lacquer is not pure white, but rather greyish or yellowish. Many white pigments, such as zinc oxide, zinc sulphide, calcium carbonate, barium carbonate, calcium sulphate, lead white, etc., turn brown to black, and no white lacquer can be obtained with them.

Red Pigments.—Vermilion and red oxide of iron. These two are used for the red lacquer, but vermilion should be stoved at a low temperature.

Blue Pigment.—Prussian blue.

Yellow Pigments.—Cadmium sulphide, lead chromate and orpiment.

Green Pigment.—Chromium oxide (? Guignet's green).

Black Pigment.—Lamp black. This is one of the pigments for black lacquer, but does not give a brilliant colour, therefore it is better to prepare the black lacquer by adding iron powder or some compound of iron to the lacquer.

Various mixed colours are obtained by mixing some of the above-mentioned pigments.

Examples of application are as follows:—

(1) *Golden Yellow.*—Finished lacquer, 10 parts; gamboge, 1 to 3; solvent, 5. If utensils are lacquered with this thin lacquer and dried for about 2 hours in an air-oven at a temperature of 120° C. a beautiful hard coating of golden colour is obtained.

(2) *Black.*—Black lacquer, 10 parts; solvent 2 to 4. Utensils lacquered with this lacquer are dried for about an hour at 130° to 140° C.

(3) *Red.*—Vermilion, 10 parts; finished lacquer, 4; solvent, 2. This lacquer is dried for about an hour at 130° to 140° C.

(4) *Khaki or Dirty Yellow.*—Barium sulphate, 100 parts; chromic oxide, 3; finished lacquer, 20 to 25; solvent, 15. This lacquer is dried for about half an hour at 160° C.

(5) *Green.*—Barium sulphate, 100 parts; chromic oxide, 20 to 50; finished lacquer, 40 to 50; solvent, 20. This is dried for about 10 minutes at 160° C.

(6) *Yellow.*—Barium sulphate, 100 parts; lead chromate, 40; finished lacquer, 40; solvent, 20. This is dried for about 15 minutes at 150° C.

Almost all pigments other than the above-mentioned are blackened by contact with lacquer or suspend its drying quality.

Several organic lakes can be used for coloured lacquers, that is to say, Indian yellow, thioflavin, and auramine lake for a yellow lacquer; fuchsine, rhodamine, and chloranisidin lake for a red; diamond sky blue, and patent nileblue lake for a blue; acid green, diamond green, brilliant milling green, vert-methyl lake, etc., for a green; methyl violet, acid violet, and magenta lake for a violet; phloxine lake for a pink. These lakes, however, are decomposed more or less on heating and fail to give proper colours when dried at a high temperature.

Modern Methods of Japanning and Enamelling with Natural Japanese Lacquer.

Urushiol, the principal constituent of Japanese lacquer, does not according to the Japanese investigator, Kisaburo Miryama, dry by itself at ordinary temperatures, but can be dried with ease at a temperature above 96° C. In the same way, lacquer that has been heated to a temperature above 70° C. and has entirely lost its drying quality can be easily dried at a high temperature. In this method of japanning the higher the temperature is, the more rapidly does the drying take place; for instance, a thin layer of urushiol, or lacquer, hardens within 5 hours at 100° C., within 30 minutes at 150° C., and within 10 minutes at 180° C. Japanning at a high temperature with natural lacquer does not require the presence of the enzymic nitrogenous matter in the lacquer, and gives a transparent coating which is quite hard and resistant to chemical and mechanical action; in these respects it is distinguished from that dried at an ordinary temperature. During the drying, oxygen is absorbed from the atmosphere and at the same time a partial decomposition takes place.

This method of japanning has its application in lacquering metal work, glass, porcelain, earthenware, canvas, papier-mâché, etc.; because the drying is affected in a short time, and the coating thus obtained is much more durable than the same obtained by the ordinary method.

For practical purposes it is better to *thin the lacquer with turpentine oil or other solvent* in order to facilitate the lacquering and lessen the drying time of the lacquer. Since the lacquer-coating turns brown at a high temperature, lacquers of a light colour should be dried at 120° to 150° C.; and even those of a deep colour must not be heated above 180° C. *Most pigments are blackened by lacquer; therefore the varieties of coloured lacquers are very limited.*

FOOTNOTES:

[1] A question has been raised concerning the safety of Perkins' apparatus, not merely as relates to the danger of explosion, but also respecting that of high temperature; and it has been asserted that the water may be so highly heated in the tubes as to endanger the charring and even inflammation of paper, wood, and other substances in their contact or vicinity: such no doubt might be the case in an apparatus expressly intended for such purposes, but in the apparatus as constructed by Perkins, with adequate dampers and safety valves, and used with common care, no such result can ensue. Paper bound round an iron tube is not affected till the temperature exceeds 400°; from 420° to 444° it becomes brown or slightly singed; sulphur does not inflame below 540°.

SECTION V.

COLOURS FOR POLISHED BRASS—MISCELLANEOUS.

Painting on Zinc or on Galvanized Iron.

Painting on zinc or galvanized iron is facilitated by employing a mordant of 1 quart of chloride of copper, 1 of nitrate of copper, and 1 of sal-ammoniac, dissolved in 64 parts of water. To thin mixture add 1 part of commercial hydrochloric acid. This is brushed over the zinc, and dries a dull-grey colour in from twelve to twenty-four hours, paint adhering perfectly to the surface thus formed.

Bronzing Compositions.

The following are the formulæ for a variety of baths, designed to impart to polished brass various colours. The brass objects are put into boiling solutions composed of different salts, and the intensity of the shade obtained is dependent upon duration of the immersion. With a solution composed of sulphate of copper, 120 grains; hydrochlorate of ammonia, 30 grains; and water 1 quart, greenish shades are obtained. With the following solution, all the shades of brown, from orange-brown to cinnamon, are obtained: chlorate of potash, 150 grains; sulphate of copper, 150 grains; and water, 1 quart. The following solution gives the brass first a rosy tint, and then colours it violet and blue: sulphate of copper, 435 grains; hyposulphite of soda, 300 grains; cream of tartar, 150 grains; and water, 1 pint. Upon adding to this solution ammoniacal sulphate of iron, 300 grains, and hyposulphite of soda, 300 grains, there are obtained, according to the duration of the immersion, yellowish, orange, rosy, and then bluish shades. Upon polarizing the ebullition, the blue tint gives way to yellow, and finally to a pretty grey. Silver, under the same circumstances, becomes very beautifully coloured. After a long ebullition in the following solution, we obtain a yellow-brown shade, and then a remarkable fire-red: chlorate of potash, 75 grains; carbonate of nickel, 30 grains; salt of nickel, 75 grains; and water, 10 oz. The following solution gives a beautiful dark-brown colour: chlorate of potash, 75 grains; salt of nickel, 150 grains; and water, 10 oz. The following gives in the first place, a red, which passes to blue, then to pale lilac, and finally to white: orpiment, 75 grains; crystallized sal-sodæ, 150 grains; and water, 10 oz. The following gives a yellow-brown: salt of nickel, 75 grains; sulphate of copper, 75 grains; chlorate of potash, 75 grains; and water, 10 oz. On mixing the following solutions, sulphur separates, and the brass becomes covered with iridescent crystallizations: (1) cream of tartar, 75 grains; sulphate of copper, 75 grains; and water, 10 oz. (2) Hyposulphite of soda, 225 grains; and water, 5 oz. Upon leaving the brass objects immersed in the following mixture, contained in corked vessels, they at length acquire a very beautiful blue colour: hepar of sulphur, 75 grains; ammonia, 75 grains; and water, 4 oz.

A Golden Varnish for Metal.

Take 2 oz. of gum sandarach, 1 oz. of litharge of gold, and 4 oz. of clarified linseed oil, which boil in a glazed earthenware vessel till the contents appear of a transparent yellow colour. This will make a good varnish for the final coating for enamelled and japanned goods.

Carriage Varnish.

The following is used for the wheels, springs, and carriage parts of coaches and other vehicles: Take of pale African copal 8 lb.; fuse, and add $2^1/_2$ gallons of clarified linseed oil; boil until very stringy, then add $1/_4$ lb. each of dry copperas and litharge; boil, and thin with $5^1/_2$ gallons of turpentine; then mix while hot with the following varnish, and immediately strain the mixture into a covered vessel. Gum anime, 8 lb.; clarified linseed oil, $2^1/_2$ gallons; $1/_4$ lb. each of dried sugar of lead and litharge; boil, and thin with $5^1/_2$ gallons of turpentine; and mix it while hot as above directed. Of course these quantities will only do for big jobs, and as it has to do with metal, it has been thought advisable to include the formula in this handbook.

Metal Polishes.

The active constituent of all metal polishes is generally chalk, rouge, or tripoli, because these produce a polish on metallic surfaces. The following recipes give good polishing soaps:—

(1) 20 to 25 lb. liquid soap is intimately mixed with about 80 lb. of Swedish chalk and $1/2$ lb. Pompeiian red. (2) 25 lb. liquid coco-nut oil soap is mixed with 2 lb. tripoli, and 1 lb. each alum, tartaric acid, and white lead. (3) 25 lb. liquid coco-nut oil soap is mixed with 5 lb. rouge and 1 lb. ammonium carbonate. (4) 24 lb. coco-nut oil are saponified with 12 lb. soda lye of 38° to 40° B., after which 3 lb. rouge, 3 lb. water, and 32 grammes ammonia are mixed in. Good recipes for polishing pomades are as follows: (1) 5 lb. lard and yellow vaseline is melted and mixed with 1 lb. fine rouge. (2) 2 lb. palm oil and 2 lb. vaseline are melted together, and then 1 lb. rouge, 400 grains tripoli, and 20 grains oxalic acid are stirred in. (3) 4 lb. fatty petroleum and 1 lb. lard are heated and mixed with 1 lb. of rouge. The polishing pomades are generally perfumed with essence of myrbane. Polishing powders are prepared as follows: (1) 4 lb. magnesium carbonate, 4 lb. chalk, and 7 lb. rouge are intimately mixed. (2) 4 lb. magnesium carbonate are mixed with 150 grains fine rouge. An excellent and harmless polishing water is prepared by shaking together 250 grains floated chalk, 1 lb. alcohol, and 20 grains ammonia. Gilded articles are most readily cleansed with a solution of 5 grains borax in 100 parts water, by means of a sponge or soft brush. The articles are then washed in pure water, and dried with a soft linen rag. Silverware is cleansed by rubbing with a solution of sodium hyposulphite.

Black Paints.

Carbon, in one form or another, is the base of all black pigments. By far the most common of these, as used in structural plants, is graphite. Other black pigments are lamp-black (including carbon black) and bone-black, the former being produced in many grades, varying in price from twopence to half a crown per pound. Bone-black, which is refuse from the sugar-house black, varies in the percentage of carbon contained, which is usually about 10 or 12 per cent, the remainder being the mineral matter originally present in the bone, and containing 3 or 4 per cent of carbonate, whilst most of the remainder is phosphate of lime. Lamp-black is an absolutely impalpable powder, which having a small amount of greasy matter in it, greatly retards the drying of the oil with which it may be mixed. For this reason it is not used by itself, but is added in small quantity to other paints, which it affects by changing their colour, and probably their durability. For example, it is a common practice to add it to red lead, in order to tone down its brilliant colour, and also to correct the tendency it has to turn white, due to the conversion of the red oxide of lead into the carbonate.

Black Stain for Iron.

For colouring iron and steel a dead black of superior appearance and permanency, the following is a good formula: 1 part bismuth chloride, 2 parts mercury bi-chloride, 1 part copper chloride, 6 parts hydrochloric acid, 5 parts alcohol, and 50 parts lamp-black, these being all well mixed. To use this preparation successfully—the article to be blacked or bronzed being first made clean and free from grease—it is applied with a swab or brush, or, better still, the object may be dipped into it; the liquid is allowed to dry on the metal, and the latter is then placed in boiling water, the temperature being maintained for half an hour. If, after this, the colour is not so dark as is desired, the operation has simply to be repeated, and the result will be found satisfactory. After obtaining the desired degree of colour, the latter is fixed, as well as much improved generally, by placing for a few minutes in a bath of boiling oil, or by coating the surface with oil, and heating the object till the oil is completely driven off The intense black obtained by this method is admirable.

Another black coating for ironwork, which is really a lacquer, is obtained by melting ozokerite, which becomes a brown resinous mass, with a melting-point at 140° F. The melted mass is then further heated to 212° F., the boiling-point of water. The objects to be lacquered are scoured clean by rubbing with dry sand, and are dipped in the melted mass. They are then allowed to drip, and the ozokerite is ignited by the objects being held over a fire. After the ozokerite has burned away, the flame is extinguished, and the iron acquires a firmly adhering black coating, which resists atmospheric influences, as well as acids and alkalies. If the black iron vessels are to contain alkaline liquids, the above operation is repeated.

A good cheap stock black paint or varnish for ironwork is prepared, as follows: Clear (solid) wood tar, 10 lb.; lamp black or mineral black, $1^1/4$ b.; oil of turpentine, $5^1/2$ quarts. The tar is first heated in a large iron pot to boiling-point, or nearly so, and the heat is continued for about 4 hours. The pot is then removed from the fire out of doors, and while still warm, and not hot, the turpentine, mixed with the black, is stirred in. If the varnish is too thick to dry quickly, add more turpentine. Benzine can be used instead of turpentine, but the results are not so good. Asphaltum is preferable to the cheap tar.

To make another good black varnish for ironwork, take 8 lb. of asphaltum and fuse it in an iron kettle, then add 2 gallons of boiled linseed oil, 1 lb. of litharge, $1/2$ lb. of sulphate of zinc (add these slowly, or the mixture will boil over), and boil them for about 3 hours. Then, add $1^1/2$ lb. of dark gum amber, and boil for 2 hours longer, or until the mass will become quite thick when cool. After this it should be thinned with turpentine to the proper consistency.

Varnishes for Ironwork.

A reliable authority gives the following as a very good recipe for ironwork varnish. Take 2 lb. of tar oil, $1/2$ lb. of pounded resin, and $1/2$ lb. of asphaltum, and dissolve together, and then mix while hot in an iron kettle, taking all care to prevent the flames getting into contact with the mixture. When cold the varnish is ready for application to outdoor ironwork. Another recipe is to take 3 lb. of powdered resin, place it in a tin or iron vessel, and add thereto $2^1/2$ pints of spirits of turpentine, which well shake, and then let it stand for a day or two, giving it an occasional shake. Then add to it 5 quarts of boiled oil, shake it thoroughly well all together, afterwards letting it stand in a warm room till it gets clear. The clear portion can then be drawn off and used, or reduced with spirits of turpentine till of the requisite consistency. For making a varnish suitable for iron patterns, take sufficient oil of turpentine for the purpose of the job in hand, and drop into it, drop by drop, some strong commercial oil of vitriol, when the acid will cause a dark syrupy precipitate in the oil of turpentine, and continue to add the drops of vitriol till the precipitate ceases to act, after which pour off the liquid and wash the syrupy mass with water, when it will be ready for use. When the iron pattern is to be varnished, it must be heated to a gentle degree, the syrupy product applied, and then the article allowed to dry.

A fine black varnish suitable for the covering of broken places in sewing machines and similar articles, where the japanned surface has become injured or scratched, can be made by taking some fine lamp-black or ivory-black, and thoroughly mixing it with copal varnish. The black must be in a very fine powder, and to mix the more readily it should be made into a pasty mass with turpentine. For the ordinary repairing shop this will be found very handy.

The following is a simple way for tarring sheet-iron pipes to prevent rusting. The sections as made should be coated with coal tar, and then filled with light wood shavings, and the latter set alight. The effect of this treatment will be to render the iron practically proof against rust for an indefinite period, rendering future painting unnecessary. It is important, of course, that the iron should not be made too hot, or kept hot for too long a time, lest the tar should be burnt off.

The following is a varnish for iron and steel given by a recognized authority: 5 parts of camphor and elemi, 15 parts of sandarach, and 10 parts of clear grains of mastic, are dissolved in the requisite quantity of alcohol, and applied cold.

Another good black enamel for small articles can be made by mixing 1 lb. of asphaltum with 1 lb. of resin in 4 lb. of tar oil, well heating the whole in an iron vessel before applying.

A good brown japan can be prepared by separately heating equal quantities of amber and asphaltum, and adding to each one-half the quantity by weight of boiled linseed oil. Both compounds are then mixed together. Copal resin may be substituted for the amber, but it is not so durable. Oil varnish made from amber is highly elastic. If it is used to protect tin-plate printing, when the plates after stoving have been subsequently rolled so as to distort the letters, the varnish has in no way suffered, and its surface remains unbroken.

A bronzing composition for coating iron consists of 120 parts mercury, 10 parts tin, 20 parts green vitriol, 120 parts water, and 15 parts hydrochloric acid of 1.2 specific gravity.

SECTION VI.

PROCESSES FOR TIN-PLATING.

In these days of making everything look what it is not, perhaps the best and cheapest substitute for silver as a white coating for table ware, culinary vessels, and the many articles requiring such a coating, is pure tin. It does not compare favourably with silver in point of hardness or wearing qualities, but it costs very much less than silver, is readily applied, and can be easily kept clean and bright. In tinning hollow ware on the inside the metal article is first thoroughly cleansed by pickling it in dilute muriatic or sulphuric acid and then scouring it with fine sand. It is then heated over a fire to about the melting-point of tin, sprinkled with powdered resin, and partly filled with melted pure grain tin covered with resin to prevent its oxidation. The vessel is then quickly turned and rolled about in every direction, so as to bring every part of the surface to be covered in contact with the molten metal. The greater part of the tin is then thrown out and the surface rubbed over with a brush of tow to equalize the coating; and if not satisfactory the operation must be repeated. The vessels usually tinned in this manner are of copper and brass, but with a little care in cleaning and manipulating, iron can also be satisfactorily tinned by this means. The vessels to be tinned must always be sufficiently hot to keep the metal contained in them thoroughly fused. This is covering by contact with melted tin.

The amalgam process is not so much used as it was formerly. It consists in applying to the clean and dry metallic surface a film of a pasty amalgam of tin with mercury, and then exposing the surface to heat, which volatilizes the latter, leaving the tin adhering to the metal.

The immersion process is the best adapted to coating articles of brass or copper. When immersed in a hot solution of tin properly prepared the metal is precipitated upon their surfaces. One of the best solutions for this purpose is the following: —

Ammonia alum $17^{1}/_{4}$ oz.

Boiling	12½ lb.
Protochloride of tin	1 oz.

The articles to be tinned must be first thoroughly cleansed, and then kept in the hot solution until properly whitened. A better result will be obtained by using the following bath, and placing the pieces in contact with a strip of clean zinc, also immersed:—

Bitartrate of potassium	14	oz.
Soft water	24	"
Protochloride of tin	1	"

It should be boiled for a few minutes before using.

The following is one of the best solutions for plating with tin by the battery process:—

Potassium pyrophosphate	12	oz.
Protochloride of tin	4½	"
Water	20	"

The anode or feeding-plate used in this bath consists of pure Banca tin. This plate is joined to the positive (copper or carbon) pole of the battery, while the work is suspended from a wire connected with the negative (zinc) pole. A moderately strong battery is required, and the work is finished by scratch-brushing.

In Weigler's process a bath is prepared by passing washed chlorine gas into a concentrated aqueous solution of stannous chloride to saturation, and expelling excess of gas by warming the solution, which is then diluted with about ten volumes of water, and filtered, if necessary. The articles to be plated are pickled in dilute sulphuric acid, and polished with fine sand and a scratch-brush, rinsed in water, loosely wound round with zinc wire or tape, and immersed in the bath for ten or fifteen minutes at ordinary temperatures. The coating is finished with the scratch-brush and whiting. By this process cast-or wrought-iron, steel, copper, brass, and lead can be tinned without a separate battery. The only disadvantage of the process is that the bath soon becomes clogged up with zinc chloride, and the tin salt must be frequently removed. In Hern's process a bath composed of —

Tartaric acid	2 oz.
Water	100 "
Soda	3 "
Protochloride of tin	3 "

is employed instead of the preceding. It requires a somewhat longer exposure to properly tin articles in this than in Weigler's bath. Either of these baths may be used with a separate battery.

SECTION VII.

GALVANIZING.

Galvanizing, as a protecting surface for large articles, such as enter into the construction of bridges, roofs, and shipwork, has not quite reached the point of appreciation that possibly the near future may award to it. Certain fallacies existed for a long time as to the relative merits of the dry or molten and the wet or electrolytical methods of galvanizing. The latter was found to be costly and slow, and the results obtained were erratic and not satisfactory, and soon gave place to the dry or molten bath process, as in practice at the present day; but the difficulty of management in connexion with large baths of molten material, and the deterioration of the bath, and other mechanical causes, limit the process to articles of comparatively small size and weight. The electro deposition of zinc has been subject to many patents, and the efforts to introduce it have been lamentable in both a mechanical and financial sense. Most authorities recommend a current density of 18 or 20 ampères per square foot of cathode surface, and aqueous solutions of zinc sulphate, acetate or chloride, ammonia, chloride or tartrate, as being the most suitable for deposition. Electrolytes made by adding caustic potash or soda to a suitable zinc salt have been found to be unworkable in practice on account of the formation of an insoluble zinc oxide on the surface of the anode and the resultant increased electrical resistance; the electrolytes are also constantly getting out of order, as more metal is taken out of the solution than could possibly be dissolved from the anodes by the chemicals set free on account of this insoluble scale or furring up of the anodes, which sometimes reaches one-eighth of an inch in thickness. To all intents and purposes the deposits obtained from acid solutions under favourable circumstances are fairly adhesive when great care has been exercised to thoroughly scale and clean the surface to be coated, which is found to be the principal difficulty in the application of any electro-chemical process for copper, lead, or tin, as well as for zinc, and that renders even the application of paint or other brush compounds to futile unless honestly complied with. Unfortunately these acid zinc coatings are of a transitory nature, Their durability being incomparable with hot galvanizing, as the deposit is porous and retains some of the acid salts, which cause a wasting of the zinc, and consequently the rusting of the iron or steel. Castings coated with acid zinc rust comparatively quickly, even when the porosity has been reduced by oxidation, aggravated no doubt by some of the corroding agents—sal-ammoniac, for instance—being forced into the pores of the metal. Other matters of serious moment in the electro-zincing process, apart from the slowness of the operation, were the uncertain nature, thickness, and extent of the coating on articles of irregular shape, and the formation of loose, dark-coloured patches on the work; the unhealthy and non-metallic look and want of brilliancy and the lustre prevented engineers and the trade from accepting the process or its results, except for the commoner articles of use. To obviate any tendency of the paint to peel off from the zinc surface, as it generally manifests a disposition to do, it is recommended to coat all the zinc surfaces, previous to painting them, with the following compound: 1 part chloride of copper, 1 part nitrate of copper, 1 part sal-ammoniac, dissolved in 61 parts of water, and then add 1 part commercial hydrochloric acid. When the zinc is brushed over with this mixture it oxidizes the surface, turns black, and dries in from twelve to twenty-four hours, and may then be painted over without any danger of peeling. Another and more quickly applied coating consists of, bi-chloride of platinum, 1 part dissolved in 10 parts of distilled water, and applied either by a brush or sponge. It oxidizes at once, turns black, and resists the weak acids, rain, and the elements generally.

Zinc surfaces, after a brief exposure to the air, become coated with a thin film of oxide—insoluble in water—which adheres tenaciously, forming a protective coating to the underlying zinc. So long as the zinc surface remains intact, the underlying metal is protected from corrosive action, but a mechanical or other injury to the zinc coating that exposes the metal beneath, in the presence of moisture causes a very rapid corrosion to be started, the galvanic action being changed from the zinc positive to zinc negative, and the iron, as the positive element in the circuit, is corroded instead of the zinc. When galvanized iron is immersed in a corrosive liquid, the zinc is attacked in preference to the iron, provided both the exposed parts of the iron and the protected parts are immersed in the liquid. The zinc has not the same protective quality when the liquid is sprinkled over the surface and remains in isolated drops. Sea air, being charged with saline matters, is very destructive to galvanized surfaces, forming a soluble chloride by its action. As zinc is one of the metals most readily attacked by acids, ordinary galvanized iron is not suitable for positions where it is to be much exposed to an atmosphere charged with acids sent into the air by some manufactories, or to the sulphuric acid fumes found in the products of combustion of rolling mills, iron, glass, and gas works, etc., and yet we see engineers of note covering-in important buildings with corrugated and other sheets of iron, and using galvanized iron tie rods, angles, and other constructive shapes in blind confidence of the protective power of the zinc coating; also in supreme indifference as to the future consequences and catastrophes that arise from their unexpected failure. The comparative inertia of lead to the chemical action of many acids has led to the contention that it should form as good, if not a better, protection of iron than zinc, but in practice it is found to be deficient as a protective coating against corrosion. A piece of lead-coated iron placed in water will show decided evidences of corrosion in twenty-four hours. This is to be attributed to the porous nature of the coating, whether it is applied by the hot or wet (acid) process. The lead does not bond to the plate as well as either of the other metals — zinc, tin, copper, or any alloys of them. The following table gives the increase in weight of different articles due to hot galvanizing: —

Description of Article	Weight of Zinc per Square foot	Percentage of Increase of Weight
Thin sheet-iron	1.196 oz.	18.2
$5/16$-in. plates	1.76 "	2.0
4-in. cut nails	2.19 "	6.72

Description of Article	Weight of Zinc per Square foot	Percentage of Increase of Weight
7/8-in. die bolt and nut	approximately 1.206 oz.	1.00

Tin is often added to the hot bath for the purpose of obtaining a smoother surface and larger facets, but it is found to shorten the life of the protective coating very considerably.

A portion of a zinc coating applied by the hot process was found to be very brittle, breaking when attempts were made to bend it; the average thickness of the coating was .015 inch. An analysis gave the following result: tin, 2.20; iron, 3.78; arsenic, a trace; zinc (by difference), 94.02. A small quantity of iron is dissolved from all the articles placed in the molten zinc bath, and a dross is formed amounting in many cases to 25 per cent of the whole amount of zinc used. The zinc-iron alloy is very brittle, and contains by analysis 6 per cent of iron, and is used to cast small art ornaments from. A hot galvanizing plant, having a bath capacity of 10 feet by 4 feet by $4^1/_2$ feet outside dimensions, and about 1 inch in thickness, will hold 28 tons of zinc. With equal amounts of zinc per unit of area, the zinc coating put on by the cold process is more resistant to the corroding action of a saturated solution of copper sulphate than is the case with steel coated by the ordinary hot galvanizing process; or, to put it in another form, articles coated by the cold process should have an equally long life under the same conditions of exposure that hot galvanized articles are exposed to, and with less zinc than would be necessary in the ordinary hot process. The hardness of a zinc surface is a matter of some importance. With this object in view aluminium has been added from a separate crucible to the molten zinc at the moment of dipping the article to be zinced, so as to form a compound surface of zinco-aluminium, and to reduce the ashes formed from the protective coverings of sal-ammoniac, fat, glycerine, etc. The addition of the aluminium also reduces the thickness of the coating applied. Cold and hot galvanized plates appear to stand abrasion equally well. Both pickling and hot galvanizing reduce the strength, distort and render brittle iron and steel wires of small sections.

The End.

www.ingramcontent.com/pod-product-compliance
Lightning Source LLC
Chambersburg PA
CBHW030506220526
45464CB00006B/2681